YOUR KNOWLEDGE HAS VALUE

- We will publish your bachelor's and
  master's thesis, essays and papers

- Your own eBook and book -
  sold worldwide in all relevant shops

- Earn money with each sale

Upload your text at www.GRIN.com
and publish for free

Nawshirwan Rashid

Effects of Establishing New Railroad on the Economic Development

GRIN Publishing

Imprint:

Copyright © 2015 GRIN Verlag, Open Publishing GmbH
Print and binding: Books on Demand GmbH, Norderstedt Germany
ISBN: 978-3-668-01070-3

This book at GRIN:

http://www.grin.com/en/e-book/302550/effects-of-establishing-new-railroad-on-the-economic-development

GRIN - Your knowledge has value

Since its foundation in 1998, GRIN has specialized in publishing academic texts by students, college teachers and other academics as e-book and printed book. The website www.grin.com is an ideal platform for presenting term papers, final papers, scientific essays, dissertations and specialist books.

Visit us on the internet:

http://www.grin.com/

http://www.facebook.com/grincom

http://www.twitter.com/grin_com

Effects of establishing new Railroad on the Economic Development

Nawshirwan Rashid

Human Development University

Collage of Economy and Administration

Department of Banking Administration

July 2014

Abstract

Today in the economic environment of each country there are some important factors which make the economy on the growth road, among these factors is the infrastructure and mobility (internally and externally), in the period of industrialization the mobility of resources, raw materials and freights was expensive because of the limitation of mobility access; the primary aim of discovering new ways and systems to an easy and cheap mobility was earning more and more benefits. Establishing new ways to transport easy and cheap was the discovering the rail ways to move passengers and freights among the cities first and the countries after, this new way of transporting made a huge distributing of the economic growth in those pioneer countries, which distributed changing public policies of more and more countries toward establishing the new transport manner, deciding to establish a new railways systems distributed to bring more investments and benefits to the investors and the productivity and growing GDP per capita of those countries.

2

Table of Content

Introduction

For generations of observers and researchers, the revolution in transportation, particularly the railroad, has appeared majorly as a determining factor for the development and settlement of different countries in the world (Levinson, 2012; Onyewuenyi, 2011; Salzberg, 2013). However, there have been notable debates as regards whether development in transportation contributed to economic growth or simply came after economic growth. While the significance of the railroad to economic growth may be determined by comparing the emerging railway systems with different economic indicators, it is important to note the possible influence of external economic variables on the development of the economy. Thus, a positive change in the economic conditions of a country during, or shortly after the introduction of railway infrastructure, does not necessarily mean that the economic improvement results from the introduction of the railway systems.

The understanding of the possible influence of other external factors, apart from railroads, on the development of the economy inspired some research studies to focus on clarifying the relationship between the railroad infrastructure and economic growth. While there is clear indication of a relationship between railroad constructions and economic development, most of these studies do not assert if the economic growth is a result of the railroad systems, or if the railroad systems were actually a product of economic growth. This paper presents utilizes case studies from both developed and developing countries, to identify the position of railroad construction on economic development. The case studies investigate the effective economic impact on the introduction of new railroad technology in developed countries, and also analyses the effects of the railroad systems on the economic growth of developing countries that initially had no railroad systems.

Introduction of New Rail Systems in Developed Regions

High-speed rail (HSRs) lines have been developed and proposed in different countries around the world. The merits of these systems is that they present an improved service quality than competing modes (such as bus, air, conventional, and auto), faster speed depending on the particular locations, shorter loading and unloading periods, improved safety, and less labor costs.

However, the disadvantage of these systems majorly lies in higher fixed and energy costs, and louder noise. Whether the benefits outweigh the costs is not the focus of this research. HSRs are usually introduced to developed countries, which have the technology and facilities required for such rail systems. This section of the paper presents an analysis of the influence of HSR technology introduction on the economy of two developed countries, USA and China.

Elhorst and Oosterhaven (2008) project direct and indirect advantages (increased customer benefits, indirect reduction congestion, spatial labor market reorganization outcomes, labor market size, and international labor outcomes) from a Maglev system proposed for the Netherlands. The indirect benefits range from 0% to 38% of the direct benefits. The results of interviews with decision-makers at corporations in Utrecht, find some corporations located close to the apparent accessibility of intercity rail connections and urban transit, while some were indifferent. Nevertheless, high-speed trains had no significant impacts on the choice of location for any of the corporations since the benefits over conventional trains were minor and connections necessitated transfers (Willigers, 2003).

Nakamura and Ueda (1989) found that three out of the six areas in Japan with Shinkansen stations had a larger population growth-rate than national average from 1980 to 1985, with no area without the Shinkansen having a higher growth rate than the national average.

It remains unclear whether the connection is that the rail led to the economic growth or regions expected to grow attracted rail investment remains unclear. Comparable research studies performed regarding metropolitan growth shows results which indicated a correlation between Shinkansen and growth (for example Hirota, 1984), however the causal structure remains unclear. Recent research studies suggest that the effects of the more recent Shinkansen lines were not as favorable as previous lines (Nakagawa & Hatoko, 2007). Sands (1993) states that the Shinkansen has expanded growth, but has not caused it.

Albalate and Bel (2010) explains that cities served by High Speed Trains (HSTs) gain from enhanced accessibility, but simultaneously there is a reduction of traditional train and air services on the lines where a HST alternatives are accessible. HSTs do not seem to attract enhanced services corporations, which indicate no greater tendency to locate in regions neighboring HST stations. Also, while business tours and conferences gain from HST services, a

5

decrease in the amount of overnight stays reduce tourist expenses and the utilization of hotel services, thus negatively influencing those aspects of the economy. Interestingly, even though HST line enhance accessibility between the metropolises they link, it disarticulates the area between these metropolises – creating what is called "tunnel effect" (Gutiérrez, et al, 2005). Thus, HST systems seem not to increase inter-territorial unity, but instead they result in territorial division.

Reviewing the influence of Europe, Japan, and other areas with HSRs, Sands (1993), predicted that HSRs in California will reinforce current population and employment and economic growth tendencies. Kim (2000) predicts that HSRs in South Korea will result in a concentration of the population within and around Seoul, while dispersing employment. Even though HSR will present an advantage for its users, 'the high investments in HST system may be unjustifiable based on the economic growth effects owing to their uncertainty (Givoni, 2006).

USA Case

The network structure of HSR lines is usually the hub-and-spoke design, connecting hub cities (e.g. Madrid, Tokyo, and Paris) to subordinate cities in tree-like design. These networks have infrequent crossing links, characteristically at lower frequency, lower speed, and lower construction costs than mainlines. Since these systems are nationally designed, with the biggest city usually serving as the capital (as in Madrid, Tokyo, and Paris), which is (approximately) centrally positioned, it is not surprise where the hub in US was located. The objective for the hub-and-spoke network is to achieve density economies in the track utilization and network outcomes at the hub cities which permit frequent service to various locations. Various paths between the origin and the destination would spread the network effects and lead to less frequent services, and thus reduce the overall demand. The hub-and-spoke pattern, while being beneficial to the network in general when demand is not sufficient to permit recurrent point-to-point service, clearly aids the hub cities the highest, as they take advantage of the incoming flows which lead to increased and therefore higher level of service. In air transport systems, airlines usually utilize the hub-and-spoke system, and if there is a large market share located at the hub airport, will utilize harness the opportunity and charge premiums for travelers, thus capturing majority of the benefits consumers get for residing in a hub city.

Hall (2009), explains that the spatial impacts of the new lines would be complicated. He explains that they will favor the major large central cities they link, particularly the urban cores of these cities, and this would in turn, threaten the situation of peripheral locations. Preston and Wall, (2008) explains that the wider economic advantages of HSRs are difficult to identify, since they are flooded with other external factors, but the economic impacts would more possibly be identified in more central settings than in the peripheral positions (Levinson, 2012). For the purpose of this report, a hub is an activity center, from which a minimum of three spokes (connections to other cities) originate.

The proposed US Program (illustrated in the figure below) has no actual national construction. There have been a number of proposals on one map. These may be viewed as hubs located at different significant points as tabulated below:

Table 1: Types of Hubs Cities and their effects on urban creative economics

Degree (Gini = 0.661)[a]		Betweenness (Gini = 0.866)		Closeness (Gini = 0.471)	
New York	26,503,070	Atlanta	18,366,780	Chicago	108
Los Angeles	21,166,460	Chicago	12,507,880	Atlanta	105
Chicago	17,271,180	Dallas	11,298,840	Detroit	102
Miami	14,896,710	Charlotte	10,086,900	Orlando	99
San Francisco	14,546,780	Denver	8,373,880	Dallas	98
Las Vegas	12,816,180	Houston	7,053,900	Houston	87
Orlando	12,681,740	Phoenix	6,053,630	New York	86
Dallas	12,006,540	Detroit	5,038,850	Minneapolis	83
Atlanta	11,469,230	Washington	3,813,680	Washington	83
Denver	11,194,100	Minneapolis	3,563,560	Denver	81

[a] Gini coefficients are computed for the complete sample (N = 128)

Source: Zachary P. Neal, Types of Hub Cities and their Effects on Urban Creative Economies, Chapter 10, communicating link https://www.msu.edu/~zpneal/publications/nealhubs.pdf

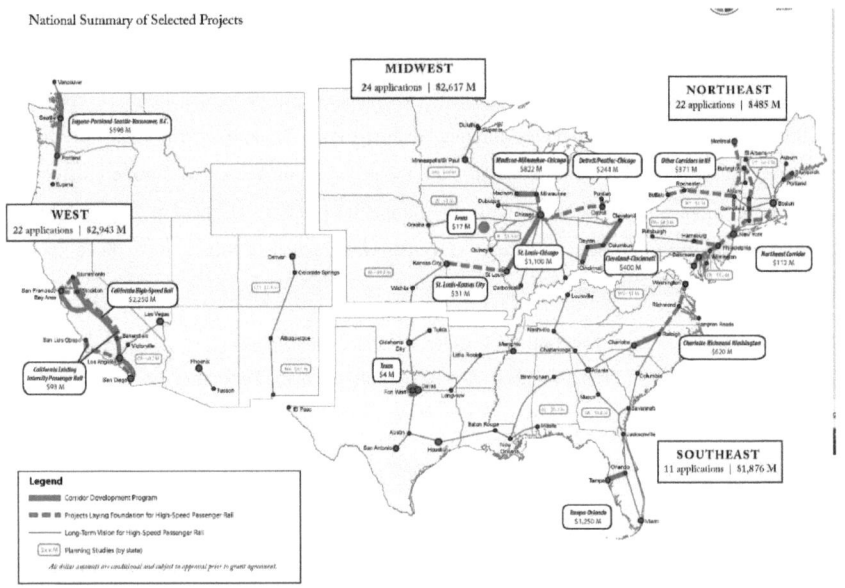

Figure 1: High-Speed Intercity Passenger Rail Program.
Source: D.M. Levinson / Journal of Transport Geography 22 (2012) 288–291

Numerous connect the different multiple hub networks (such as New Orleans, Raleigh, Louisville and Kansas City).

The political strategic advantage of the proposed intercity passenger network is that includes in 42 states.3 This is an important strategy learned in transportation from one initial national systems, the Interstate Highway System (which lines in all fifty 50 states, as well as special routes in Hawaii and Alaska). There are two functions of accessibility; the first is to increase the overall wealth. Agglomeration economies influenced by new infrastructure increases aggregate output (Levinson, 2012). Secondly however, it redeploys wealth, as the destinations with higher accessibility gain more of the aggregate wealth than the destinations with a lower aggregate gain. Places that do not increase accessibility as high as the average may end up losing economic opportunities that naturally occur at areas with accessibility advantages.

8

Hubs, owing to their respective locations, will capture accessibility advantages disproportionate to the already considerably large population share they have. First beneficiaries are Chicago, New York, Seattle, Orlando, and Los Angeles, since they serve as the HSR network hubs. Locations where the network links will also experience some advantages, but not as much as the hub cities. Second tier beneficiaries are Dallas and Atlanta, which are second generation hubs. Third tier beneficiaries include cities such as Kansas, New Orleans, Raleigh and Louisville which connect numerous hubs. The other cities on the networks will also experience absolute accessibility advantages, with people in such cities being able to connect with more people in a shorter period (or with a better, or with lower out-of-pocket expenses). Nevertheless, while they may accomplish absolute advantages in accessibility, they may forfeit in comparative position, as a higher share of the currently larger total availability becomes occupied by the hub cities.

There is yet no actual empirical research study directly relaxing the economic growth impacts of HSR in the US. A limited amount of research has been written from objective studies. Peterman et al., (2009) made efforts to balance the objectivity through the Congressional Research Service. The employment estimates from California would be huge if approved validated. One infrastructure project was resulting in 450,000 employment opportunities (for an overall civilian employment of less than 16 million5) given a total of approximately 3% of the workforce of the state. Construction related employments alone are 1% of the workface of the state. Currently, construction is approximately at 577,000 employments, so the project has tendency to absorb on about a third or a quarter of all construction projects in the state (Levinson, 2012).

Even though the publicity of project agents may not be reasonable, there are logical regional results. It is possible to argue about improving the strength of the intercity connections to modify the current municipal system into the megalopolitan method, in which people more frequently interrelate between cities. If, based on Adam Smith's suggestion, the labor division is restricted by the level of the market, and transportation may be utilized to increase the market, the labor division may then increase (i.e. become more specialized), which will have more economic benefits (similar to economy of agglomeration). Melo et al. (2009) performs a meta-analysis of approximations of urban agglomeration economies using 34 research studies. The results show a considerably large range of effects, with no clear conclusions regarding the magnitudes that may be drawn. The researchers' findings "support the intuition that agglomeration estimates for any particular empirical context may have little relevance elsewhere" (Levinson, 2012). On the possibility that HSR will increase markets is dependent on if it has a faster (point-to-point) than substitute methods of transportation, and if it will increase the productivity of its users. These possibilities are both on the context.

This section of the paper reviewed the state of HSR planning in the US. The proposals generally request for a collection of barely inter-related hub-and-spoke interactions. Various points may be raised. Based on the results of the research in the US case, evidence from research studies indicate that lines have two basic influences. There is the positive accessibility advantage in metro regions serviced served by stations, however there are also negative nuisance influences. HSLs are not likely to have local accessibility advantages apart from the role they play in linking local transit lines since there is negligible benefits for most individuals or businesses to locate close to a line which is utilized infrequently. Nevertheless, they may have more extensive metropolitan level influences. They will sustain, and possibly, have more nuisance results than domestic transit. An earlier research on the full costs of HDR in California by Levinson et al. (1997) indicated that the vibration and noise along the line may be important, even if there is an installation of noise reduction technologies.

The domestic land use influences around HSR stations would possibly be negligible. Agglomeration advantages may be present, but there is limited basis for deciding their size. If HSLs can produce bigger effective regions, then they may influence the distribution of who benefits and losses from such infrastructure. The amount of agglomeration economies remains uncertain (and undoubtedly location-specific), however, it is likely that it presents the optimal case that may be to support the US HSR system. The introduction HSR systems in the US therefor indicate the possibility of influencing some economic variables, such as employment, increased efficiency of business service delivery.

Railroad and Economic Impact in Regions Initially without Railroad Systems

The two case studies reviewed in the previous section of the paper, one for a developed country (USA) and one for a somewhat developed region (Guangzhou) within a developing country (China), it is suggestive that the situation of railroads is influenced by the need to connect already existing economic hubs. However, it will be difficult to identify the effects of railway on the economy of the tributary regions and/or location were the railroads pass through without identifying a case study of the construction of a railway system in a country which railway systems were initially non-existent. This section of the paper review a case of Nigeria, a country that initially derived its major economic stability through agriculture until the discovery of oil deposits. Nigeria's case highlights the relationship between the construction of the railway network and the changes in the performance of agricultural produce during, and shortly after, the construction of the railways.

China Case: Railroad Expansion in Developing Countries

There are contradictions at the core of transport project assessment. Policy-makers, especially in developing regions, are likely to perceive major transport system investments via a lens that expresses their influence on regional economic growth, expanding market accessibility for products and services, and distributing ideas and skills. Nevertheless, project evaluation practitioners have more scarcely concentrated on evaluating the direct travel expenses and benefits, which is, the worth of operator cost saving and commuter travel time savings, including drops in externalities including accidents, noise pollution, an d air pollution. They have considered that the broader economic influences are too vague to be consistently measured and/or are merely a consequent of direct advantages. Inspire of the obvious advantages of taking a careful view particularly when large quantities of public funds are involved, there is increasing consensus amongst economists that most transport investments could important effects that are not properly documented through these regular cost-benefit examinations (Fujita, Venables, & Krugman, 1999).

The illogicality is also obvious in China's improvement of high speed rail (HSR), where provincial and national governments insist on the significance of regional economic growth in practically all feasibility analysis, in spite of the absence of these ideas in quantified evaluations

11

of economic contributions (Salzberg, Bullock, Jin & Fang, 2013). Their ideologies are based on, partially, on evidence accrued in other circumstances. Research studies have indicated that transport enhancement may inspire economic activity if they are able to materially enhance accessibility for personal and business commuters – specifically, improving connections to major regional and national information and commerce hubs. It is most likely that a new business located within everyday reach of these hubs will be more reachable to a wider labor pool and other business activities, thus increasing productivity (UK Department for Transport, 2006). This connections is a significant element of what are widely called agglomeration economies, the advantage that accrue to organizations and people from the crowding of economic happenings.

Effectively measuring these advantages depends on an ideology referred to as 'Economic Mass'. The economic mass is an evaluation that integrates the size of the economy of a city with the accessibility of the city to other locations. Therefore, the economic mass in a particular region may therefore improve via one of two ways: the rate economic activity may grow, or the neighboring regions may become increasingly accessible, as evaluate by an integration of the travel cost and time. Thus, if HSR (or other enhancements in transport) may cut down the travel friction between regions, it may enhance the 'economic masses' of the regions it aids.

The ideology and evaluation of economic mass is considerably straightforward; where studies have advanced in recent years has remained in observing the connection between a region's economic mass and the overall productivity level. This connection is based on four suggestions: (1) economic mass incases with improvement in transportation; (2) average employee outputs, and their wages, differ directly the increase in economic mass, even when control is made for other variables; (3) positive externalities exists from transportation enhancements which lead to an increase in the output for some corporations independently of their transport use networks; and (4) the increase in output is not part of the standard assessment of transportation projects. (Salzberg, et al. 2013)

One major research study made efforts to quantify this conation and approximated that, under normal circumstances, a 100% increase in economic mass would result in an increase in per employee productivity (Salzberg, et al. 2013). Crucially, the productivity advantages accrue to individuals and businesses even if they do not travel. Majority of the research studies discussed so far in this section focus on developed economies, with few making efforts to investigate the parameters that guide these connections in developing regions. The World Bank performed two different research studies in order to bridge bias of research studies concentrating majorly on developed countries. One of the studies presents a statistical analysis connecting economic mass and productivity in southern China's Guangdong region (Salzberg, et al. 2013). The other study presents a case of agglomeration as it may take place in reality in a regional midpoint in China.

Amongst the three mega-city areas of China, The Province of Guangdong appears to have the highest prospects for such a research study considering that the region is China's most productive and contributes the highest provincial share (contributing more than 10%) of the country's GDP. Guangdong's current spatial growth is also informative. Guangdong hosted three Special Economic Zones (SEZs) selected in 1979 by China. Since then, Guangdong SEZs recorded markedly varying growth courses. Shenzhen, existing adjacently to Hong Kong and therefore directly exposed to the major economic mass, succeeded. It had a yearly average GDP growth of 15% from 2000 to 2008. Zhuhai, following Macau but somewhat distant from the major economic midpoints, has recorded an annual growth rate of 13% within the same period. The third SEZ, Shantou, is situated over 450km from Guangzhou and was not linked with the national expressway system until 2003, remains the region with the lowest GDP growth rate (at 9% annually). Even though these trends are mere anecdotes on their own, the present opportunity for future research studies.

Although the research studies mentioned above have connected economic mass and productivity in most developed regions, China is yet to be classified as a developed country. Despite Guangdong, being amongst the wealthiest of China's provinces, it has a per capita GDP of USD 6,500 only in 2008, which in reality is equal to the US output level of 1930.

Basic and manufacturing industries, typically labor intensive and low-tech, were responsible for more than 70% of the output of Guangdong. The high-end Research and Development business services sectors signified only a small percentage of the tertiary segment results. Empirical proof for developed countries may not be directly transferrable to other parts of China, including Guangdong.

In order to determine the extent and nature agglomeration in China, it would be necessary to consider the relationship between productivity and economic mass using within a specific period using urban and county district level economic statistics. One of such studies focused on the relationship between productivity and economic mass from 2005 to 2008.

The study results showed that economic mass proximity is typically linked with a higher rate of average earnings and this positive association is continually robust after controlling for other variables including academic level, industry composition and capital investment. Generally, the best approximations of the flexibility of the productivity regards to economic mass are more than twice the economy-wide outcomes observed in comparable research studies performed for developed countries. The optimal models in the estimations so far indicate that increasing the economic mass would 100% will result in a 9-15% increase in per employee productivity. This is more than the general suggestion that such an increase result to productivity growth of 5-8% (Rosenthal & Strange, 2004).

It is safe to utilize a 5.3% growth to evaluate the influence of HSR projects in China. Theories and econometric approximation would only offer a partial understanding of the complicated economic influence of transport enhancement as they play out in reality. Case studies would expose the actual influences of the choices that influence the economic reactions of businesses in bordering regions newly linked with major economic areas. Yunfu is a town on the border of Guangdong's Pearl River Delta. It takes about approximately two and a half hours during non-peak times and the major urban section is not connected by rail.

The HSR line between Nanning and Guangzhou will open is proposed to open in December, 2013 and will significantly reduce the total travel time from 2.5 hours to 40 minutes. Currently, Yunfu has the least GDP rating in Guangdong, with a 9.8% yearly average GDP growth rate from 2000 to 2008. In the course of interviews, local ventures agreed that attempts to attract investment, skills, and growth opportunities are hindered by poor transport accessibility. Yunfu therefore offers a representative environment for assessing the development influence of transport systems through a pre and post study.

14

It is possible that HSR services may influence Yunfu's economy in different ways: opportunity for innovative ideas and the entrance of or accessibility to highly skilled employees; net capital and labor inflow to Yunfu; expansion of export-based secondary ventures with enhanced productivity; and a general boost to distinctive service industry subsectors, including recreation and tourism. None of these impacts are guaranteed. Enhanced accessibility may entice local investors and skill to Shenzhen and Guangzhou, therefore wearying domestic businesses. There is also a possibility of differential influences within the town based on the significantly domesticated ease of access to the introduced HSR.

The overall economic influence of HSR will be dependent on the degree to which the domestic occupants anticipate and harness of the new transportation development by modifying business strategies and operations, adapting intensity and patterns of urban land usage, and intensifying entrepreneurial and social interactions. Results of a field trip suggest that the domestic business municipal is considerably innovative and that urban development policies and plans have already begun to change for the future HSR linkage. While the changes in business approaches resulting from the proposal of the HSR system indicates the influence of railroad systems on the economic variables of business (such as new strategies) It will be necessary to conduct a follow up research-study when the NanGuang line is opened so as to identify the eventual effects.

Based on the just highlighted study, a team of World Bank researchers in China have started piloting a methodological approach to evaluate the agglomeration advantages for the HSR project. A good example is the examination for a new 200km/hr. high speed passenger railway between Nanning and Guangzhou. The new route connecting the already existing railway system of a region in southwest China to the main part of the Pearl River Delta, saving over 220 km in travel length. It more directly connects the Pearl River Delta with two of the country's poorest provinces (Yunnan and Guangxi) and railway services to other inland cities located around the Pearl River presently served by neither expressway nor railway. A significant policy objective of the project is to offer support to the regions in reducing the discrepancy in economic conditions of the residents and the wealthier coastal residents of the country.

Survey of businesses and planners in Nanning indicted that local businesses well informed of the NanGuang Railway construction. Nanning businesses and along as well as the

NanGuang corridor have also created their commercial efforts to harness the complementarity of local business activities to businesses in the Pearl River Delta. Also, Guangdong based businesses have also started made considerable investments in Nanning businesses, attracting skill as well as investment for growing and startup business activities. Upon completion, the rail travel time from Nanning to Guangzhou for commuters will reduce from the present eleven hours to about three hours, offering a quantum enhancement in the accessibility of Yunnan and Guangxi to Pearl River Delta's commerce and industry.

The case study in this section indicates the presence of agglomeration economies in China. The case also makes it obvious that Chinese HSR projects can deliver notable agglomeration advantages in the correct context. Considering the relatively short travel periods for which HSR is most beneficial, making sure that stations are suitably located for the residents and businesses in the city is vital. This includes integrating HSR service both using urban development strategies and with other transport methods.

The Nigeria Case: Railway and Agricultural Growth

The British colonial leadership in Nigeria realized that the more efficient the areas of supply and demand could be connected, the less important distances would be from the economic standpoint. The British specifically relied majorly on rail systems to form the important connection between the internal and external ports.

One major advocate of railway utilization in Nigeria's economic of growth was Sir F. D. Lugard. Discussing the potential of trade in the Nigeria's Northern region, Lugard (1906) wroter:

> "...until better transport is arranged and the Iron horse takes the place of human carriers, Northern Nigeria cannot possibly realise the hopes and wishes of those who are assured of the great future before it." Again, referring to the Zungeru-Barijuko Light Railway that was started a year before, he argued that the extension of this line would "render possible the export of cotton and other produce grown in the Nupe Province and in Southern Zaria" (Northern Nigeria, 1902, p. 58).

The insufficiency of the then river and road networks for the requirements of the country resulted in the need to construct railway lines. Nigeria's railway network development ensued in a way comparable with the general rules of Morrill, Gould, and Taaffe. The earliest three phases of the model are of specific significance when discussing Nigeria's railway network development particularly in terms of its economic contributions, since they are identifiable in Nigeria's railway network evolution. Initially, the Nigerian waterways and ports, and adjacent lands served as the means for contact. Ports situated at the banks and estuaries of major waterways emerged as the trade centers. The previous railway penetration lines emanated from the ports located at Baro, Lagos, Zungeru, and Port Harcourt.

The development of the earliest Nigerian railway started in in December, 1898, named the Lagos Government Railway, during which a few hundred meters of railway lines were laid. Gaining northwards via Abeokuta the lines arrived Ibadan (197 km), which was the then center for cocoa-production, in 1901 and Jebba, in Niger state (494 km) by 1909. The first railway to be constructed in the North was the 35km Zungeru-Barijuko Light railway, commenced in 1901 at Zungeru and was finished in 1902 at Kaduna (Barijuko). The major railway line which penetrated the Northern part of Nigeria was, the Baro-Kano Railway, a 573 km, which construction commenced in 1908 at Baro on *The Niger River* bank, and was finished in 1911 at its Kano terminal, which was then the center of the groundnut region through Zaria and Kaduna. The tin mining parts of Jos were connected in 1914 with the railway through the narrow gauge 227km Bauchi Light Railway.

In 1913, the Eastern Line constructions commenced from Port Harcourt. In 1916, having passed 61km through Aba, the line arrived at the Enugu coal mines (243 km) by 1916. The figure below is an illustration of the first phase of the routes of Nigeria's railway system (Fig.2)

The next phase of railway construction directly connected more of the internal centers to the ports, and was more completely achieved in 1915 during the connection of the Lagos and Baro-Kano lines at the Jebba bridge, and also in 1927 after the Eastern lines stretched from Kaduna, to Enugu and Jos. From 1896 and 1927 the two major north-south connections were formed therefore providing remote hinterland accessibility to the two major ports of Port Harcourt and Lagos. The next phase of the network development was achieved in the Nigerian railway from 1927 to 1930 through the contrsuction of the extensions and branch lines including:

Kano-Nguru (229 km), Zaria-Kaura Namoda (232 km), Kafanchan-Jos, and Ifo-Idogo (39 km). Towards the end of 1930, each of the railway lines were laid except the 639 km Bukuru-Maiduguri extension which was finished in 1964.

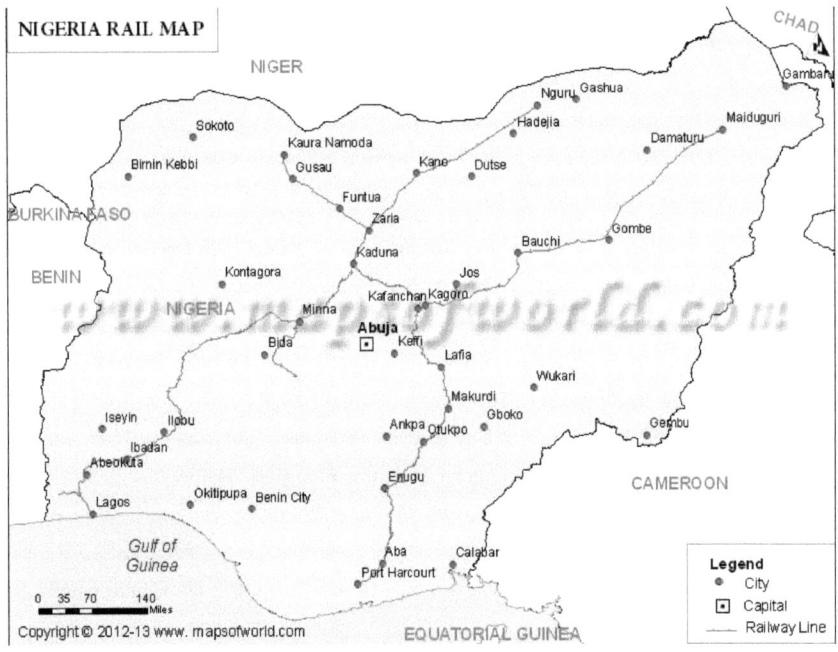

Figure 2: Map of Nigeria Rail
Source: Map of world Current, Credible, Consistent http://www.mapsofworld.com/nigeria/rail-map.html

Related with this transportation development stage was the growth of emerging centers of economic activities. During of integration periods of the railways system in 1912, the total numbers of stations open to traffic were about 84. The amount of station increased with the increment of the route length and 1922 became 138, 216 by 1932 and 218 by the end 1950. When the main lines were completed in 1930, more than 200 buying and selling stations had begun along the railway line routes. The most significant stations were "Lagos, Abeokuta, Ibadan, Ilorin, Minna, Bara, Kaduna, Zaria, Kano, Nguru, Funtua, Gusau, Kaura Namoda, Jos,

18

Makurdi, Enugu, Aba and Port Harcourt" (Onyewuenyi, 2011). It is important to mention Port Harcourt specially it started existing only in 1913 when its location was selected as the terminus for the new Eastern Line. From then it has emerged into the second biggest port in the Nigeria, surpassing other older established ports such as Degema, Opobo, and Bonny.

The role of the Nigerian railway system was majorly for the goods services. Out of the 3,930 units of rolling stock functioning in 1934, 3,605 were dedicated to goods service with only 325 actually dedicated to coaching services. By 1913, coaching services were responsible for 18.6% of gross revenue, with goods services accounting for 79.5%. The influence of goods services increased to 81.6% by 1931 and 86.2% by 1934. 64% of the goods traffic in 1902 comprised of agricultural produce (Nigerian Railway, 1933). This position remained practically unchanged all through the period averaging 64.4% annually from 1931 to 1950 (as indicated in the table below Table 2). Therefore agricultural products transported kept pace with the general development of the railway traffic system.

Table 2: Railway Goods Traffic: 1931-32, 1964-47 and 1949-50

Year		Total Goods (Paying) Traffic	Total Agricultural Products (Traffic)	% Agricultural to Total Goods
1931	32	581,094	340,912	58.7
1932	33	613,550	395,521	64.5
1933	34	599,546	394,036	65.7
1934	35	624,445	421,538	67.5
1935	36	709,102	429,680	60.6
1936	37	891,848	556,939	62.4
1946	47	967,206	660,726	68.3
1947	48	892,676	580,726	65.1
1948	49	1,024,896	660,068	64.4
1949	50	1,045,968	696,016	66.5
				X = 64.4

Source: **Adapted from Annual Reports on the Nigerian Railway. THE EFFECT OF RAILWAY CONSTRUCTION ON THE GROWTH OF EXPORT , AGRICULTURE: THE NIGERIAN EXPERIENCE. Remy N Onyewuenyi, CSSp., Ph. D. Faculty of Environmental Sciences, Caritas University, Amorji-Nike, Enugu

Although it is true that Nigeria's railway system has been influential in stimulating crop product exportation, the reaction of the various crops to and their dependency on the system greatly varied. The fluctuating reaction was mainly a result of the crop's nature, the spatial arrangement of its producing region, and level of demand for the crops. After the extension of the railway to the production regions of annual crops such as cotton, groundnut, and sheanut, the growth in output was observable within eighteen months. For a few tree crops such as rubber, oil palms, increased harvesting ensued almost instantly with the establishment of rail lines and feeder access roads. The reaction was much later from other crops like cocoa. Usually, Cocoa required 4 to 7 years to start producing. The Ibadan (the Cocoa producing city in Nigeria) case is a notable example of the delayed reaction. The railway arrived Ibadan early in 1901 and the Cocoa output started showing considerable market growth in 1908. Market value cost of transportation and the production location areas were also related factors to the reaction of various cash crops to the railway system. Northern cash crop production seems to have reacted more quickie and favorably to the than the southern export crops, since the railways afforded the much required direct links with Nigeria's coastal areas and external markets, which till then were partially accomplished by river transportation.

Table 3: Percentage Quantity of Groundnut, Palm Produce, Cocoa, and Cotton Exports transported by Rail, between 1931-32 and 1949-50

Year		Groundnuts	Palm Kernel	Palm Oil	Cocoa	Cotton
1931	32	84.0	15.6	25.5	34.1	54.2
1932	33	97.0	24.2	23.2	37.3	67.8
1933	34	84.0	15.9	29.7	44.9	79.3
1934	35	89.8	19.3	19.9	37.9	80.2
1935	36	89.4	22.3	24.9	40.2	60.5
1936	37	88.7	18.7	27.4	34.6	92.9
1937	38	87.6	21.7	25.2	36.7	80.7
1946	47	98.0	17.5	42.8	11.9	91.2
1947	48	95.6	16.2	43.3	8.0	95.9
1948	49	134.4*	15.3	40.9	3.8	92.3
1949	50	102.8*	10.3	36.9	.9	99.0
Averages		95.6	16.6	28.4	26.4	81.3

Sources: **Adapted from Annual Reports on the Nigerian Railway; Colonial Reports on Nigeria; Department of Statistics (Trade Reports) for the various years. THE EFFECT OF RAILWAY CONSTRUCTION ON THE GROWTH OF EXPORT , AGRICULTURE: THE NIGERIAN EXPERIENCE. Remy N Onyewuenyi, CSSp., Ph. D. Faculty of Environmental Sciences, Caritas University, Amorji-Nike, Enugu

From 1902 to 1950, oil palm produce recorded the lowest mean percentage (22.5%). The considerably low amount of railed palm produce was a result of some factors: (1) the palm belt travels the East to West direction through the rain-forest zone a region which does not allow extensive railroad construction due to its dense vegetation and numerous creeks and rivers. The railway passes from North to South direction passing through only a fraction of the production areas (2) the road networks and the river offered optional competitive transportation options. The railway merchandise rates for short hauls were more in most situations (See Table 4), (5) the bulking of palm kernels, particularly, in sacks or bags made them easily transportable through simple means. Various bags were easily transported to the coasts, at once using bicycles.

Table 4: Railway Freight Rates: Agricultural Exports, 1952-53

Commodity	Distance	Rate per Ton	Rate per Ton km (pence)
Cocoa	193 km (Ibadan-Lagos)	32/6	2.03
Cotton. Ginned	1159 km (Gusau-Lagos)	132/3	1.37
Groundnuts	1127 km (Zaria & North to the ports)	182/3	1.95
Groundnuts	560 km (Zaria & North to Baro)	142/9	3.08
Hides & Skins	1357 km (Nguru-Lagos)	228/-	2.03
Kola-nuts	1084 km (Ifo-Kano)	305/-	3.39
Palm Oil	61 km (Aba-Port Harcourt	20/-	3.95
Yams	161 km	16/-	1.20
Yams	805 km	28/6	0.43

Source: Adapted from International Bank for Reconstruction and Development, X=2.16

1955, p.469.

THE EFFECT OF RAILWAY CONSTRUCTION ON THE GROWTH OF EXPORT , AGRICULTURE: THE NIGERIAN EXPERIENCE. Remy N Onyewuenyi, CSSp., Ph. D. Faculty of Environmental Sciences, Caritas University, Amorji-Nike, Enugu

The groundnut exportations from Kano area alone indicated an increase to the tune of 1,536% in less than, from 1911, prior to the railway arrival in Kano, to 1913, one year after the railway arrived. The increase is attributable to the extending of the railway into areas of production. Responses of some of the newspaper media then highlighted a casual association between the railway arrival and increased trade and production.

Africa Mail's March issue reported thus:

21

> *"A really quite extraordinary rush of traffic has taken the railway by surprise at Kano, and although it was fully expected that Kano trade would develop yet it was hardly anticipated that during what is practically the first full season of the line being really open for traffic, a trade in groundnuts alone would tax the resources of the railway"* (Morel, 1913, p. 238)

Lamb (1913) also stated that the:

> *"The arrival of the railway at Kano has within eighteen months given tremendous stimulus to groundnut cultivation in the neighbourhood . . . Every available piece of land is being planted with groundnuts . . ."* (p. 663).

The growth noted in groundnut production, specifically in 1926, 1929 and 1933, are connected to the expansion of plantation areas in the eastern parts of Sokoto province where trade was opened by the expansion of railways including Bauchi, Gusau and Funtua, (1926) and Namoda, and Kaura (1929), and where it feeder roads expanded trades; as well as the new regions which were steadily improved through the Nguru Branch Line (1930). Considering the percentage railed it appears as through railway was responsible for influencing majority of the production growth experienced in the course of this period.

Some cotton quantities, approximately 11 tons, were exported early in the early 1900s. It was after 1906 subsequent to the construction of the roads and railways into the Western Region of Nigeria, that exportation stared assuming significance. The introduction of the railway connection to the Northern part of Nigeria in 1912, amongst other factors, appeared to have positively influenced cotton exportation. By 1916, a volume of more than 3,300 tons of cotton were achieved, a 135% increase of what was achieved in 1906. Additional increases were reported during the 1930s. The increases were linked with the extension of the railroads hinterland and more specifically owing to the change in production regions. As at this time, the North was emerging as a major manufacturer of marketable cotton. The change in production regions tool place majorly owing to the production of low quality cotton in the western parts of the country, and the optimal quality cotton brought into Nigeria by British Cotton Growers

22

Association, subsequently in the century, which adapted better to country's Northern geological pattern. With the drop in the demand for cotton in the western parts of the country, western farmers substituted cotton farming for cocoa, which was even more lucrative. Approximately 81 percent of cotton exportation was railed to the ports from 1931 to 1950.

According to the railed proportion, one may assume a high level of dependence of cotton transportation on the railway system. With a delay in the commencement of the cotton and groundnut exportation until 1912, it is possibly not an overstatement to assert that the provision of railway transport remains the single most significant factor in Northern Nigeria's economic growth.

Railways majorly influenced the cultivation and exportation of cocoa up until the early periods of the 1940s when there was a decline in shipment from 37% to 0.9% between 1934 and 1950. The reduction was a result of the intense level competition and apparent takeover of every short distance rail available by road, and owing to that apart from Oshogbo and Ibadan, and the Abeokuta area, the rail evaded the major production areas. The rails were fed by feeder road networks. As at 1921, a facility of motor lorries linking the towns of Oyo, Ibadan, Oshogbo and Ogbomosho, and in the midst of Ilesha and Oshogbo, were operating (Colonial Report, Nigeria, 1921). However, since World War II the railway has progressively become an auxiliary to road transportation systems in the cocoa production regions.

Based on some of the crops railed, one may compute the estimate the estate under particular crops in the economic vicinity of some of the railway stations. It is important to note that this form of measurement is not the ideal approach for determining the level of land-use or hinterland limitations of any specific area. When there is no total agricultural production and actual land-use data, the railway exportation data are the optimal option. The level of generalization and interpretation is restricted to what the data permits. Notwithstanding the limitations, some overall land use patterns may be recognized through the approximations for the period prior to the 1940s. Acreage under cultivation grew at all station regions, with more increases presenting at the hub areas. For example, acreage under cotton increased from a yearly average of 84,700 in for 1933 and 1934 to 244,600 for 1935 and 1937. Even though there was a reduction in the planted in cotton in the course of this period 1947 to 1950, the cultivated area was still roughly 29% above the 1932 and 1934 period (Table 5).

Table 5: Estimate* of Acreage under Cotton: Railway Station Areas, 1923-33 to 1949-50

Stations	1932-33 1933-34	1934-35 1936-37	1947-48 1949-50
Lafenwa	284	765	179
Ilugun	-	1,152	16
Adio	-	1,725	-
Ibadan	3,758	27,955	135
Iwo	737	2,569	-
Ede	996	4,742	21
Oshogbo	2,002	8,604	114
Zaria	9,838	31,823	15,146
Gimi Dabosa	2,040	3,979	299
Challowa	-	4,847	12
Kano	885	1,996	1,244
Saba	1,297	2,282	423
Duchi-N-Wai	8,469	10,801	2,285
Kudaru	-	4,550	2,298
Rahama	1,922	6,797	1,621
Funtua	27,755	51,869	44,478
Gusau	15,933	671,972	28,456
Kaura Namoda	3,222	6,451	2,657
Jos	-	-	1,963
Zungeru	-	-	2,148
Total	84,711	244,587	109,260

* Estimates are based on the annual averages of crop output moved by rail. The figures are calculated by dividing the quantity of crop railed by estimated yield per acre.

Source: **Adapted from Annual Reports on the Nigerian Railways for the various years.
THE EFFECT OF RAILWAY CONSTRUCTION ON THE GROWTH OF EXPORT , AGRICULTURE:
THE NIGERIAN EXPERIENCE. Remy N Onyewuenyi, CSSp., Ph. D. Faculty of Environmental Sciences,
Caritas University, Amorji-Nike, Enugu

The case of Nigeria presents an important review of the relationship between railway systems and economic growth. From the review, it is apparent the economic indicators may have influenced the situation of railway systems however, the influence of railway systems on already existing economic hubs, as well as its instigation of other initially dormant economic activities and regions, shows that railway systems did more for the Nigerian economy than the economy did for it.

Time series chart indicating railway and economic growth in Nigeria

Figure 3: Performance of Rail Transport and Economic Growth in Nigeria (1970-2011).

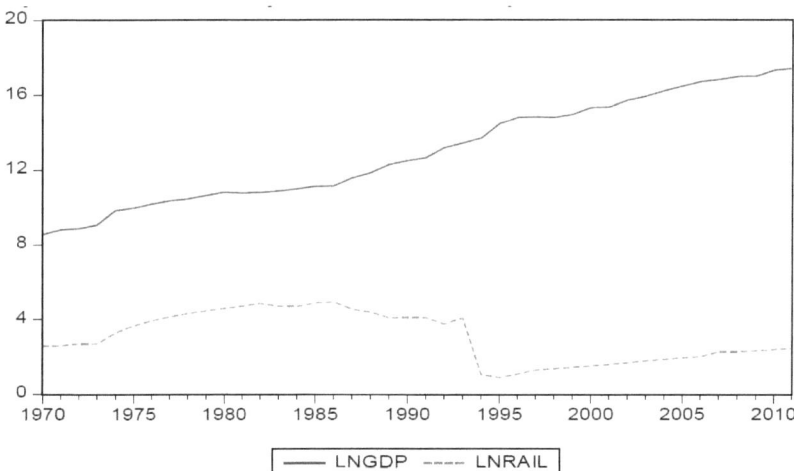

Source: APANISILE Olumuyiwa Tolulope, Rail Transport and Economic Growth in Nigeria (1970 – 2011), Australian Journal of Business and Management Research, Vol.3 No.05 [18-24] | August-2013.

The possible influence of the railway development of Nigeria's economy after the considerably apparent period may be performed by evaluating the development of the rail system of Nigeria in the post-colonial era. While it has been established that the rail system instigated the economic development, it is still necessary to understand if the decline in the rail system had any effects on the economic development of the country. This chart above presents a time series analysis of the relationship between the railroad systems and Nigeria's economic development after Nigeria gained its independence in 1960. The red trend line in the chart represents the development of the rail system while the blue trend line represents Nigeria's economic growth. It is seen that the blue trend line continues to indicate growth in spite of the fact that the red trend line fluctuates. The outcome of the analysis indicates little or no relationship between railway performance and economic growth.

Conclusion

The aim of this paper was to highlight the relationship between railway systems and economic development. From the review of different literature focusing on this topic, it is apparent that railway systems and economic development are related. In spite of the identification of this relationship, it remains uncertain if which of the variables is actually responsible for the existence or development of the other. That is, although the correlation between railways and economic development is apparent, there is uncertainty whether railways systems affects or instigate economic growth, or whether economic growth is a product of the existence of railway systems.

Owing to the practicality of the research subject, it was important to apply case studies related to the variables, railway systems and economic growth. The research focused on HSR in US and China (Yuku) and Nigeria's railway system. The findings of the research indicate that even though railway systems are influenced by the economic growth of a specific region, the areas these lines pass through to link the developed economies, also experience significance economic development Thus, economic development is influenced by, and also influences, railway system – thus creating a mutual relationship

References

Albalate, D., Bel, G., 2010. "High-Speed Rail: Lessons for Policy Makers from Experiences Abroad." *IREA Working Papers.*

Brotchie, J., 1991. "Fast Rail Networks and Socio-economic Impacts". *Cities of the 21st Century: New Technologies and Spatial Systems,* pp. 25–37.

Elhorst, J., Oosterhaven, J., 2008. "Integral cost-benefit analysis of Maglev projects under market imperfections." *Journal of Transport and Land Use* 1 (1), 65–87.

Fujita, Venables, and Krugman (1999). *The Spatial Economy: Cities, Regions and International Trade.* Cambridge, MA: MIT Press

Givoni, M., 2006. "Development and impact of the modern high-speed train: a review." *Transport Reviews 26* (5), 593–611.

Gutiérrez Puebla, J., Garcı́a Palomares, J., 2005. "Cambios en la movilidad en elárea metropolitana de Madrid: el creciente uso del transporte privado." *In: Anales de geografı́a de la Universidad Complutense.* No. 25, pp. 331–350.

Hall, P., 2009. "Magic carpets and seamless webs: opportunities and constraints for high-speed trains in Europe." *Built Environment 35 (1),* 59–69.

Hirota, R., 1984. "Present situation and effects of the Shinkansen." In: *Presented at International Seminar on High-Speed Trains.*

Kim, K., 2000. "High-speed rail developments and spatial restructuring: a case study of the Capital region in South Korea." *Cities 17* (4), 251–262.

Levinson, D. M. (2012), Accessibility impacts of high-speed rail, Journal of Transport Geography, 22 (2012) pp. 288–291

Levinson, D., Mathieu, J., Gillen, D., Kanafani, A., 1997. "The full cost of high-speed rail: an engineering approach." *The Annals of Regional Science* 31 (2), 189–215.

Melo, P., Graham, D., Noland, R., 2009. "A meta-analysis of estimates of urban agglomeration economies." *Regional Science and Urban Economics* 39 (3), 332–342.

Nakagawa, D., Hatoko, M., 2007. "Reevaluation of Japanese high-speed rail construction: recent situation of the north corridor Shinkansen and its way to completion." *Transport Policy 14* (2), 150–164.

Nakamura, H., Ueda, T., 1989. "The impacts of the Shinkansen on regional development." *In: The Fifth World Conference on Transport Research,* Yokohama, vol. 3.

Onyewuenyi, R. N. (2011), "The Effect of Railway Construction on the Growth of Export Agriculture: The Nigerian Experience," *Madonna International of Research* vol.4 No. 1

Peterman, D., Frittelli, J., Mallett, W., 2009. "High Speed Rail (HSR) in the United States. Tech." Rep., Library of Congress, *Washington DC, Congressional Research Service.*

Preston, J., Wall, G., 2008. "The ex-ante and ex-post economic and social impacts of the Introduction of high-speed trains in south east England." *Planning Practice and Research* 23 (3), 403–422.

Rosenthal, S and WC Strange (2004). "Evidence on the nature and sources of agglomeration." *Review of Economics and Statistics,* Vol 85, pp377-393.

Salzberg, A., Bullock, R., Jin, Y. & Fang, W. (2013), "High-Speed Rail, Regional Economics, and Urban Development in China", *World Bank Office,* Beijing, China Transport Topics No. 08, pp. 1-8, http://www-wds.worldbank.org/external/default/WDSContentServer/WDSP/IB/2013/01/16/00035616 1_20130116164534/Rendered/PDF/NonAsciiFileName0.pdf

Sands, B., 1993. "The development effects of high-speed rail stations and implications for California." *Built Environment* 19 (3), 257–284.

UK Department for Transport (2006). *Transport, Wider Economic Benefits, and Impacts on GDP.* London: UK Department for Transport.

Willigers, J., 2003. "High-speed Railway Developments and Corporate Location Decisions: The Role of Accessibility." *Paper presented at the 43rd ERSA Congress Jyvaskyla,* August 27–30, 2003.

** Table: 2- 3 4 sources, THE EFFECT OF RAILWAY CONSTRUCTION ON THE GROWTH OF EXPORT , AGRICULTURE: THE NIGERIAN EXPERIENCE. Remy N Onyewuenyi, CSSp., Ph. D. Faculty of Environmental Sciences, Caritas University, Amorji-Nike, Enugu, Email: dozieonyewuenyi@yahoo.co.uk